About the author

Graham Andrews is a professional writer and science writer living on the South Coast of New South Wales. He has worked as a scientific editor, technical writer, freelance writer and writing tutor.

By the same author

A Guide to Wrought Iron and Welding

You're On Air

Easy Guide to Creative Writing

Easy Guide to Writing Winning Essays

In Your New Image

Island of the Barking Dog

Dad Kept Bees

Reach For the Sky

Practical Arc Welding

Easy Guide to Science and Technical Writing

A guide to writing articles for peer-reviewed journals

Graham J Andrews

Flairnet

First published 2014

Copyright © 2014 Graham J Andrews

ISBN   978-0-9924642-2-6

Published by Flairnet
www.flairnet.com.au
Post Office Box 645
Narooma   NSW   2546
Australia

National Library of Australia Cataloguing-in-Publication entry

Author: Andrews, Graham J., author.

Title: Easy guide to science and technical writing : a guide to writing articles for peer-reviewed journals /
Graham J Andrews.

ISBN: 9780992464226 (paperback)

Notes: Includes index.

Subjects: Technical writing--Handbooks, manuals, etc

Communication of technical information.

Communication in science.

Dewey Number: 808.066

Contact the author:

Website: www.grahamandrews.com
Email: graham@grahamandrews.com

Contents

Introduction

This book in the Easy Guide series gives the reader the basic skills and information to write well when contributing to peer-reviewed scientific or technical journals. All the information needed is contained within this book—how to write well, how to edit your work so that it is easy to read, how to structure your articles for submission to journals, and what to include and, probably more importantly, what to leave out.

This book, like the others in the series, does not depend on endless exercises. The author considers many such exercises a waste of time, as it is most unlikely that any book, no matter how massive, will contain just the right exercises that a writer will need in writing a particular piece of writing. Instead, it concentrates on the writer's project—the actual article at hand, and helps the reader fine tune the result to meet the rigid standards applied to publishing in peer-reviewed journals. This book also assumes a certain degree of understanding of the English language—the normal grammar rules and so on.

This book shows you how to structure your article. There is ample guidance given in planning your article, breaking the material down by using headings, and the details that go into each section of your article—Introduction, Methods, Results, Discussion, Conclusion and so on. There is a guide to reviewers and their role in publication—what they will be asked to consider, and the process your article will go through before acceptance by a journal, as well as information about using references and compiling lists of references.

Editing is essential to any writing, and this book gives the basics of what to do in this stage of writing. There is a comprehensive checklist so you can ensure your article should be much closer to that goal—acceptance for publication in a peer-reviewed journal.

There is useful information about the use of illustrations, graphs and tables, and how to submit the data used to compile the graphs.

Oh, how hard it is to sit at our computers and begin writing. There are those usual distractions because we don't know where to begin. There is more research to attend to, more conferences to attend, more meetings, more discussions with the university hierarchy. The cat needs patting ... the dog wants to go for a walk ... the washing up hasn't been done ... Anyway, our task is just too much. It's almost insurmountable. That manuscript is yet several years away.

So let's start putting the words down.

Something that often prevents the first words from coming is the intimidating thought of the sheer volume of words that have to be written. For instance, an article might require around 2000 words. Do you believe yourself capable of writing this much material, even on a topic of research you know so well? Probably not, and your screen will remain blank for a long, long time if you approach the task at hand in its entirety. With the use of headings—and lots of them—and sub-headings, your task immediately becomes easier. You don't write 2000 words about one subject, but perhaps 200 words about 10 related topics, or even 100 words about 20 related topics. These small units, added one to the other, will soon become a full-length article. How long have you thought about starting that article you feel so strongly about? A year? More than that?

Seeing yourself making progress with your work as each 'unit' is written, can make the difference between the struggle and the triumph. If you get stuck with one section, don't stop, work on another section that you feel more confident about. You will be in a better position to smooth any bumps and cracks when the larger part of the structure is in place. And towards the end of the large project, if you have made substantial progress, those difficult pieces you left out will seem to just fall into place. You will (or, at least, should) have the confidence now to add those hard bits.

Headings will form an outline of your proposed manuscript.

Sub-headings will make the task of writing more organised and easier. These can be removed later if you no longer need them. But in the early stages of writing, they can be arranged and rearranged many times so the logical flow becomes apparent. Also they serve as a guide to even the work out over a number of points rather than concentrating on one point more than the others.

Introduction

An outline is merely a plan. Headings are the themes of topics and carry much of the supporting detail. The outline also suggests a tentative paragraph structure.

Get each segment of your article to blend with those before it and after it. Get into the habit of using linking or transition phrases. Link a paragraph to what has immediately preceded it. Join paragraphs where possible by implying 'which leads me to say ...' Such phrases can occur at or near the beginning of a paragraph or at the end of the previous paragraph—the last sentence can introduce the next angle of the article. You can also link paragraphs by words showing logical relationships, such as 'therefore', 'however', 'even so'. Any transition should shift readers easily from one topic to the next without jolting them.

And good writing, clear writing, is the secret to getting your article published, and giving your career a push along.

This book will be suitable for anyone wanting to write for peer-reviewed science or technical journals.

Chapter 1 Getting Started

Publication in a reputable, peer-reviewed journal should be the goal of all researchers, as publication is the main means of disseminating the results or findings of what can often amount to years of research.

Career advancement these days is, to a large extent, dependent upon publication in peer-reviewed journals. It's the applicant with an impressive list of published journal articles who will be ahead of those with a meagre portfolio of published work, or none at all to their credit. This book will teach you the art of scientific and technical writing. It will teach you how to format your article, its structure, the use of references, tables, graphics, and many of the important elements of submitting an article to a scientific or refereed journal.

The communication of scientific information is valued in our society. Technical information needs to be made readily available for market promotion. Results of clinical studies have to be conveyed to doctors and healthcare workers so they, and through them, their patients, can benefit from another's work. A botanist needs to report on a recent discovery—big or small—that will help other researchers in their work. Decision makers consider new findings in relevant fields. Government decisions are often based on reports of a finding or a recommendation. The examples go on.

WHAT IS TECHNICAL WRITING?

Technical writing, by definition, can include all writing that deals effectively with subjects within science and technology.

Technical writing usually contains fewer words in each paragraph, shorter sentences, and more paragraphs. And simple words. Good technical writing requires the choice of words that convey the exact meanings intended. This means that clichés are immediately abolished from all scientific and technical writing. Jargon is kept to a minimum except when its use is unavoidable.

Yet in some ways, technical writing is little different from writing for popular magazines, apart from a format that must be adhered to. As with all writing, without clarity the writing becomes nothing but a jumble of unintelligible words decipherable only to the writer. Just because you are writing mainly for professional biologists does not mean that your style should be so heavy that any interested person other than biologists cannot understand much of what you have written. These days, much of what is technical or scientific is taken up and popularised by the media through television programs aimed at the general public, newspaper reports of medical research, atmospheric deterioration, and so on.

Technical writing should be precise, concise and accurate. No such writing can get away from the use of technical terms. But a feature of good technical writing is the use of easily understood words. The term photosynthesis must be used—there is no way to avoid its use in many botanical articles. But that does not mean that the rest of the article must be full of terms that only specialists can interpret. Too many words that a reader does not understand, means a reader lost. If necessary, it might be sensible (in some types of technical writing) to include a glossary, where technical terms are defined. (A glossary is a list of terms, so the heading needs only be 'glossary', not a 'glossary of terms').

The first sentence should be of literary brilliance. It should grab the reader's attention so he or she wants to read on. It is at this early stage that your readers have the opportunity to ask themselves: what's this all about? Is it interesting? Should I bother to read it? A bored reader would much rather do something else if your article is dull and the style flat.

Clever writing is not text that confounds even the experts. Clever writing is text that is easily accessible to any intended reader. If your intended reader cannot understand your article, there is nothing smart about what you have written.

An author should always aim for high standards of writing, in producing precise and tight writing. Because of the often professional nature of the readers, the standards of scientific or technical writing are generally higher than those set for the popular press.

Chapter 2 Good and Bad Articles

Let's start with the really bad article. It is not hard to identify these. It is easy to write them, too.

A really bad article will have been hurriedly written, with no regard to the mood or capacity of the author at the time of writing, or his tiredness at the end of the day. It will have no plan, no structure. Jottings will appear in the order that the author thought of the points to include. The author will have failed to have edited it carefully.

There are writers (but few published writers!) who write like this. Is it surprising that they remain unpublished? It is not that they do not perform brilliant research work. It is not that they do not know their subject. They are often experts in their field. It is that they do not know how to write for publication, or do not care about writing. The results are the same. They don't get their work published.

A good (as in a well-written) article will appeal to a wide readership. Good writing is work that is accessible to all interested readers. Readers should be able to grasp its message the first time, without suffering laborious pains as they re-read many sentences to see what the message is about.

Through good writing, you should be able to reach all those readers who want to read your article. There will be many readers who will adopt a closed mind policy and convince themselves that because it is scientific, they, almost by definition, would not understand what has been written. Unfortunately, it is difficult to open closed minds. Such people will not be readers you have lost, even with good writing, but they should be considered as readers you never had.

Peer-reviewed published articles are likely to be read by scientists and students—some of whom may not agree, not because it is 'scientific', but because they follow another line of thought.

In writing, it is important to write with enthusiasm for your subject. If you lack enthusiasm, imagine how uninspired your readers will be if they realise you are fulfilling a duty, rather than pursuing a desire to convey information to them.

Chapter 3 Important Issues

In this section, we will consider the Why, what, how and who?

Before you even begin to write, ask yourself a question: why am I writing this? If you are not sure, put the idea aside for a while until you have a definite purpose in writing your report or article for the journal. Too much is written and submitted to journals, when the writer does not have a clear idea of what he or she is trying to communicate to others. Words without purpose won't convey much to any reader.

Okay, you have decided that you really need to communicate through words. Ask yourself the next important question: 'what'. Answering this question will help you focus on your real purpose of writing.

'How' is the next important question. Naturally, you will reply 'through words', but consider the arrangement of your proposed article. How are you going to give your message or information to readers? In what order should the points be considered? Can you condense some of the data you have through the use of tables and charts? Will two illustrations replace one thousand words? Will one table convey even more meaning than if you tried to pass on the same information in text? Will one or two line graphs show a trend more clearly than a lot of words? Quite often the answer will be yes, as long as the graphs are created clearly—again, with the reader in mind.

Now ask yourself 'who'. Whom am I writing this for? What is my reader's background in this subject? What is their level of understanding? What is their level of education? Too often writing is carried out with complete disregard to the reader—perhaps the most important person in the whole process of communicating scientific and technical information.

THE READER

A reader of technical writing can be anyone. It can be government planners. It can include administrators who have a high level of education. It can include students, and scientists with a similar background to yourself. Readers could be well-educated people with degrees or backgrounds in a discipline that is different from your own. And it can be the interested public—those who may not possess a formal degree in a particular discipline, but who are highly motivated and well-read in several sciences, and who have a tremendous understanding of your topic.

COPYRIGHT

Most, if not all, journals, require that copyright be transferred to that journal. If this is a requirement, then make sure you enclose the necessary forms transferring copyright to that journal. Occasionally a journal will have a copyright form that in turn gives the author a licence to use the article in any manner he or she wants to. But such licences returning use of copyright back to the author are rare indeed.

MORE THAN ONE AUTHOR

Sometimes it is necessary for more than one person to write an article. This often arises when one author prepares the figures, another writes the text, while another perhaps interprets the statistical data. In listing the names of more than one author, only those who have made a genuine contribution to an article should be included, but I realise that in my saying this, disputes will break out! Authorship credit should be based only on substantial contributions to the concept and design, or analysis and interpretation of data, and to drafting the article or revising it critically for important intellectual content. The assistance of others who may have helped in a lesser way should be listed and thanked in the acknowledgements.

PATIENTS' PRIVACY

Particularly with biomedical articles, it is essential to de-identify all patients. There is nothing more disconcerting to a person to read a report in an article on communicable diseases, which is picked up by the popular press, to the effect that the index case was believed to be a 23-year-old woman who returned to her home in Cootamundra, Australia on Thai Airlines last Friday, 13th March from a five-day holiday in Bangkok. How many residents in that small country town would be guessing who had the disease? All information that could possibly identify such a person must be removed from the article. However, in de-identifying the patient's details, false data should never be substituted. De-identifying personal data may not be easy, but it MUST be done.

REVIEWERS

All technical manuscripts submitted to journals go through a review process. A manuscript is received. The editor reads it and considers if it could be suitable for the journal. If it is not, it is returned, sometimes with a note saying why it is being returned, but often the note will say that the article is not suitable.

If the article is considered to be possibly suitable at that stage, it will be sent to reviewers—sometimes one, usually two, sometimes three—who are experts in the subject matter of the article.

Reviewers are an important part of scientific publishing. They will often be asked to address certain points in relation to articles sent to them.

They might be asked about:

- Validity: this would include the thoroughness of the literature cited, the validity of the assumptions made by the authors, the methods used in the research, the case studies, statistical analysis and the conclusions;

- Relevance: the relevance of the article to the journal's readers, and its appropriateness to the journal;

- Clarity: is the message clear?

- Conciseness of the article: reviewers might be asked to comment on whether the article is written in a concise style;

- Duplication of material, unnecessary diagrams and tables, irrelevant data will be commented on, and whether the authors have written down to their readers (assumed the readers don't know much about the subject);

- Style: reviewers might be asked to comment on the style of writing, and whether it conforms to acceptable scientific style and format.

Keep these points in mind even before you begin to write your article, and you will have overcome major reasons for rejection of numerous scientific articles submitted to journals.

Chapter 4 Getting Started

Ask a writer to write five thousand words. The first question will be—about what? The next question will be—for whom? Then—why? This is where planning is probably the best tool you will have to get you started and overcome what many second-class writers call writers block.

An outline does not restrict you, but helps you clarify your thoughts on your subject, and helps to guide your thinking. As you write, you may be inclined to add more points, or modify those you have included.

Having developed an outline (the plan), many authors start with the easiest parts. These are often the methods (they know what they did), followed by the results (they know what the outcome of the study was).

It is sometimes beneficial to prepare tables and diagrams before writing any text. Tables will hold a lot of statistical data on which an author can call. Graphs, for example, clarify trends and will give a clear indication of what the results were.

In your mind, do not view the whole manuscript in its entirety. Mentally break it down into smaller parts. Writing 5000 words can be a rather daunting experience. Writing 500 words about one aspect of your study can be an easier task to grasp. Even if you break down each section with sub-headings (these can be removed later if they are not required) this can make the writing of each section even easier—writing in smaller blocks removes that barrier between getting started and never finishing the work.

In the first draft stage, don't worry too much about grammar, punctuation or spelling. All these can be corrected later. The main thing is to get your points down and then rework the draft.

The topic of each paragraph is the main idea, or theme—what the paragraph is about. The first sentence introduces the topic of that paragraph, the subsequent sentences should refine the topic and fill in the details.

Keep the flow of words going smoothly. If you stumble on a section that is causing problems, skip over that part and move on to the next section that you can write about.

Often, writers find that if they stumble for quite a while on one part of an article, there may be a good reason for the hesitation—data might not seem quite right, the method might have a flaw in it.

Bear in mind the word count in the first draft stage, and try to exceed it by even as much as twenty percent. Why? Because it is easier to delete a lot of words when you rework your article than it is to edit your work using the same number of words. Or worse still, to find those missing 300 words when you think you have finished your article.

WRITERS GUIDELINES

When preparing articles, if you follow strictly the guidelines of each journal you intend writing for, then your article is less likely to be returned to you because of style. It may, however, not fit the requirements of many journals because of a number of other reasons, such as the article is not appropriate, it is badly written, or it lacks depth or information, or for many other reasons.

In writing for a journal, you must follow the instructions to authors in the journal as to what topics are suitable for that journal, and the types of articles that may be submitted, such as original articles, reviews or case reports. You should also become familiar with other requirements for submitting your work to any particular journal. Some require one copy of the manuscript, others need three or four copies. Take note of the acceptable length of article preferred, and the particular styles used by that journal.

Some guidelines published in journals might be general, others will be specific. No matter what they are, follow them carefully.

It is obvious that some authors have not even bothered to read such guidelines when their manuscript varies significantly from the requirements detailed in the guidelines. But it is surprising how often that happens! It means that the selection of the journal was ill-considered, and possibly was just a random selection, or, worse still, possibly that the author wrote the article, and was submitting the same manuscript to a string of journals in turn. In the real world, selecting an appropriate journal this way does not work. If you have the audacity to submit your article to a journal you don't even know, don't expect your work to be accepted. While the shotgun approach might work occasionally, often it creates a lot of extra work—for you, the author, and for the editor who has to write to you when returning your work.

Often editor's comments or editorials (most journals will have them) give readers and prospective authors an idea of the types of articles the editor is seeking and is interested in receiving. And commenting on previous editorials in your covering letter ('As you mentioned in your Editorial, Vol. 64, Number 17 of March 17, ...') indicates you have had the sense to at least read the journal and to assess its suitability for your proposed article.

It is important also to remember that even if a journal does accept an article immediately, then it may, and it has the right to, make editorial changes to suit the particular style of that journal.

PRIOR PUBLICATION

Do not submit your article, even in a slightly different form, to another journal while the original remains under consideration with a journal. Nor should you submit an article to a journal on a topic you have already had published in another journal. However, an article that has been presented at a conference but not previously published might be acceptable to a journal. Check with the editor first!

Translation and publication in languages other than English do not fit strictly within these restrictions, but if you intend publishing in a language other than English, liaise with the editor of the main journal you have submitted your material to.

When submitting an article to a journal, it is always courteous—and wise—to let the editor know about its previously published status. Even if an article is based on previously published work, then the editor has a right to know. A breach of these conditions will probably preclude you from ever having an article published in that journal, and perhaps a few others as well. News travels fast in the publishing industry. And when publishing is important, don't take chances on this issue. Even an announcement of the results of your research in the media, such as a discussion on television or in an article in a newspaper before your main article has been published, will be viewed very dimly. With so much—

particularly your career—depending on scientific publishing, don't take chances with editors. It's not that editors are difficult to deal with, it's just the whole general ethics associated with scientific publishing that determine the rules.

PLANNING

Good writing requires good planning. Consider what goes into your report or article. More importantly, consider what to leave out. Only the information that is immediately relevant should be included. To be effective, the order of what is said in an article requires structural soundness and logical progression. Think about this for a moment. Logical progression. This means that the final article should read like ... this led to this ... that led to that ... that made us conclude ... and so forth. Consider the logical progress of your article as if it were conversation. If, in telling someone about what you did, or concluded from your study, you had to add '... I meant to tell you this ...' and then a bit further on you would add '... which reminds me, I should have mentioned ...' then it is time to reconsider the logical progression. This, if it does creep into the article, must be taken out in the final edits.

You must have a plan, no matter how simple that might be. With a plan, it is difficult to leave out a piece of vital information. Without a plan, you might think of that important fact when you see your article in print, and say to yourself, 'oops! I should have said something else as well that I have only now thought of.'

It is easier to rearrange a plan of a proposed report or article, than to rewrite the article and try to incorporate some sort of logical flow later.

A plan should form the basis of your report or article. It can be laid out as a series of headings and sub-headings. Rearrange the main headings so they are in a logical and sensible sequence. Then, under each of those main headings, rearrange each of the sub-headings into a logical sequence. Writing your article around these numerous headings will ensure flow only in the right direction, and progress your work easily. It will also make writing much easier, as it will be a matter of adding relevant facts to each of the numerous sub-headings. In the planning stage, it does not matter how many headings you include—at this stage, the more the better. Headings, and sub-headings, when they no longer serve a purpose, can be removed in the final stages of the writing. If on reflection a heading or sub-heading seems out of place no matter where you place it, justify its existence. If you can't, take it out.

SUB-HEADINGS

Sub-headings serve two purposes in an article. They break up an otherwise very large body of text, making it easier for the reader to concentrate on the message. They also make the task of writing the article much easier.

Sub-headings can represent different levels of the writing, so that each major section in, say, Methods, flows in a logical manner.

While some aspects of research reports differ from the general descriptive article (review articles), they are nevertheless no more difficult to write than other forms of articles. Again, the simple, logical approach to their construction is required.

Chapter 5 Sections of an Article

BIOMEDICAL ARTICLES

For biomedical journals, there is a uniform requirement for manuscripts submitted to those journals.

This type of article requires these sections:

Abstract

Introduction

Methods

Materials

Results

Discussion

Conclusion

Acknowledgements

References

There is a reason for this order. A research-type article is more likely to be followed, and understood by readers, if the research or the study being written about is described in a logical sequence—that is, in the way the study or research was conceived, designed, what was done, the results obtained by following those methods, an analysis of the results, and what you, the author, concluded from the results of the study or research.

Occasionally articles will have different headings to those mentioned, but the sequence will be similar, even if different titles are used to describe each section. For example, instead of Introduction, it might be more appropriate to call this section Background.

Keep the sections separate. Don't, for example, combine Methods and Results. They are clearly two different aspects of the study. Likewise, don't combine Results and Discussion. Again, they are clearly separate entities, and should be treated as such. Editors receiving articles that combine such categories will know that writers have failed to read the guidelines, they are amateurs at writing for journals, and will no doubt treat them as such.

TITLE

You will begin working with a title (commonly called the working title). Ensure that the title of your article is relevant to the intended content, theme or treatment of your subject.

The title you began with (the working title) might be different from the one you think is more appropriate once you have written your article (you've probably got a better feel for what you wanted to say). The original title would no doubt have served to help you focus on what you intended to write.

Unlike those used a hundred years ago, titles these days should be short and effective—the shorter the better. They should accurately tell the reader what the article is about. If a title can be written so it fits on only one line of the journal, aim for that. Be frugal with words—the fewer the better, consistent with getting the message across. If the study refers to one organism, or one region, include this information in the title such as 'The Role of *Staphylococcus* in Causing Ulcers in East Gippsland, Victoria'.

ABSTRACT

The abstract is more important than many authors think, and it must be written last, even though it appears at the beginning of an article. That's because the abstract has to be a true summary of the article. A reader can scan the abstract—a short summary of the entire report or article—to see if the article is relevant to their interests before reading the rest of the article. If the abstract is relevant, the reader can obtain and read the rest of the article. Many articles these days are indexed under their abstracts. There are numerous annual compilations of abstracts, such as Chemical Abstracts, and Biological Abstracts.

Generally an abstract is no more than 250 words (or as long as is determined by the journal). It should give the purpose of the study or the investigation, the basic procedures, such as the selection of study subjects or laboratory animals, observational and analytical methods, main findings—here it is important to include specific data and their statistical significance if possible—and the main conclusions.

An abstract should never contain material that is not presented in the main text of the article. Don't use abbreviations, unless they are well recognised, such as DNA, DDT, CSIRO. The less commonly used term dsDNA (double stranded DNA) might need clarifying.

No text in the abstract is referenced.

The abstract is generally written as one long paragraph, perhaps on some occasions, as two paragraphs, and not in the usual literary fashion required of other sections of the article.

Some journals require the authors to suggest up to about ten key words. These help in indexing the article and for running searches in databases.

INTRODUCTION

The introduction stresses the purpose of the investigation, or a setting for the report or article. The introduction also stimulates the reader's interest and provides background information that helps put your research, and your article, into perspective.

Here is what should be included in the Introduction:

- Background to the topic—what is known about the topic, or is believed about it, what is not yet known (if relevant), reports of recent relevant studies, and the importance of the subject.

- A statement of the reason why you conducted the research—such as to determine the importance of vaccinating small children against measles.

- The approach you took to answer this question.

In this section, state the purpose of the article and summarise the rationale for the study or the observation. Do not include data or conclusions from the work being reported in your particular article. Keep the writing flowing in only one direction.

METHODS

This is the section of your article that will tell readers what you did, and how you did it, so that, if necessary, they will be able to repeat the procedure you followed. There should be sufficient detail for anyone to read the article and complete the same type of investigation or study that you did.

Here is what should be included in the Methods section:

- Outline of the study design—what you did and how you did it;

- Your subjects — who or what were they, how were they chosen, why were they included, why others were not included, and your sample size;

- Any controls you used to test your hypothesis against, how were controls selected, why others were excluded;

- Details of the procedures you used, and exactly what you did to whom and to what;

- Data analysis, particularly how you derived variables, how you summarised your data, what software you used, and statistical tests you applied.

MATERIALS

Detail the materials that you used to achieve the results you got, and why you chose particular equipment or ingredients.

Let us look at some of these points in more detail.

In this section, you will describe your selection of the observational or experimental subjects, whether they were human patients or laboratory animals. If you use controls in a study (as a benchmark against which variables can be measured), these too should be detailed. All relevant details, including the ages of patients, their sex, or proportion of males and females, must be included. Any other factors relevant to the selection of controls is important—health status, education, ethnicity for a start.

In this section of the article, it will be necessary for you to identify particular methods you used, any particular apparatus that might be important—sufficient details must be provided for another reader to carry out the same study that you did, by following your written instructions.

If you used established statistical methods, say so. If methods have been used in previous studies that were different from yours, mention those and reference the literature that relates to those studies. For your article, it might be important to describe the limitations of previous studies, and to give reasons why your study is more conclusive, or more important.

If doses are used in trials, mention the doses, the intervals between their administration, the period of administration, the ingredient, the brand name of the substance, and how it was administered.

Did the ethics committee know about your experiment? It is important in research, if human or animal subjects were used, to include which ethics committee approved your research work. Generally there will be national ethics committees, or ethics committees belonging to various research institutions. Their clearance is essential before work can begin on humans or animals, and particularly so if you ever hope to get your results published.

It is not sufficient to merely present a collection of data. It will be necessary for you to include a description of the methods you used to interpret the statistical side of your investigation. Any standard methods, including statistical methods you used, should be accurately referenced.

Quantify your findings and present them with appropriate indicators of measurement error or uncertainty, such as confidence levels. Avoid having to rely only on statistical hypothesis testing, such as using P values.

Always include numbers of observations. If your study encountered losses, either through deaths of subjects, or patients moving away, or not turning up, include this information, as these numbers may affect the outcome of your results in other readers' opinions.

The reason why some patients dropped out could be important. Why most animals in your study died could be just as important.

If you use terms such as random, normal, significant, correlations or sample, define these. To some readers, such terms may have meanings different from those you are used to.

RESULTS

Present your results in logical sequence in the text, tables and illustrations. Do not repeat in the text any of the data that is included in the tables or illustrations. It is necessary to emphasise or summarise only important observations.

When you summarise the data in the Results section, specify the statistical methods you used to analyse them.

It is possible you will have collected more data than you need for the article. Your study may have changed during its course, but you nevertheless continued collecting data that you felt could be useful for some further study. Decide now what to include, and definitely what to leave out. If some data bear little relevance to your findings, don't include them.

How are you going to present the data? Clear tables and graphs can convey much to the reader. They use only a small amount of journal space, they are easy to analyse, and visually reinforce what you have written.

Always remember that data is not the same as results. Data will not stand alone. Each needs explanation.

DISCUSSION

The Discussion should include:

- An indication of the originality of your work;
- Explanations of how your findings agree with previous findings, disagree with other results, any limitations of your study that may be significant;
- The importance of your work;
- Recommendations for future studies.

In this section, emphasise the new and important aspects of your study and the conclusions that follow from them. Do not repeat in detail data or other material given in the Introduction or the Results sections of your article. Include in the Discussion section the implications of the findings and their limitations. It is usual to relate observations to those of other relevant studies.

The Discussion answers the questions posed in the Introduction, and explains how the answers fit in with what is already known about the subject. If you did not ask a question in the Introduction, do not answer it in the Discussion. There is no place for irrelevancies! Here, you can express your interpretations, your opinions, the value you place on some aspect of the findings, and possibly suggestions for future research (remember that your article can influence other researchers who may want to take your study further).

Your findings may be in disagreement with previous findings. Do not be alarmed if this is so. If your findings are valid and your research methods sound, it is possible your results will differ from previous findings. But it is necessary to detail them, and explain why you think they might be different. If the unexpected finding is important, do not play it down. Say that such results were quite unexpected, and perhaps offer an explanation.

You may have found shortcomings in your methods. If they were significant, discuss these deficiencies in this section. Say why they became shortcomings. Say what should be done about them, and what, if any, influence these could have made on the results you obtained. But don't apologise for any shortcomings unless they were a serious oversight on your part.

Where your results may not have been definitive statements, use terms such as 'these data suggest ...', or 'we imply that ...'

If you can see limitations or likely questions that others might ask after they have read your article, address them to prevent the reviewers asking them—concerns you will have to address anyway. You can thus minimise criticism in the journal after your article has been published, or passed by reviewers.

CONCLUSION

On completion of the previous sections, you should write the conclusion. This is a summary of the main sections, and an interpretation of your findings and what they mean. The Conclusion gives you the opportunity to make recommendations based on these findings. You could also, if appropriate, state what actions should be taken as a result of these recommendations. Include mention of wider implications for your findings. Link the conclusions with the goals of the study, but make sure you avoid unqualified statements and conclusions that might not be completely supported by your data or observations.

Do not draw a conclusion from too few data. Don't rely on extrapolating curves. If you can't reasonably qualify a claim, don't make it. If you speculate, tell your readers that you are only speculating. And in your Conclusion, do not show any prejudice. Keep the work honest, right to the very end. Do not be influenced by your preconceived ideas, and do not omit data because you would like to think otherwise about something that is showing up.

ACKNOWLEDGEMENTS

This is the section of your article in which you can include the names of those who contributed in a small way, but who did not warrant having their names included as authors. Acknowledgement can be made to individuals for all sorts of relevant help: checking drafts of manuscripts, financial support so you could carry out the study. But don't overdo the acknowledgements—only those who contributed something real should be included.

Disputes can erupt during the writing of an article. If this is so with your article, get written approval from those persons whom you include in the acknowledgements, otherwise readers might get the impression they endorse what you have written, when in reality they disagreed with your points of view (and hence caused the dispute).

Chapter 6 References

References! These can be painful. They needn't be, though. References are merely to show where the information you have quoted came from so readers can learn more about the subject for themselves if they are interested. If they don't agree that some respected author would have dared to have said something you claim he did, then they will want to check up the source for themselves. This is all the more reason to make sure that you quoted them correctly. Prove some readers wrong if you can, otherwise some will take great delight in trying to prove you wrong on some claim!

There are two main systems of references used (but some journals and organisations have developed their own 'home' style for references). Journals will often stipulate the referencing system they use.

The first is the Harvard system. In the text, when referring to a passage written by an author, simply put in brackets the name and year of that author (Bloggs, 1992). You will need to make sure that you list his work in the reference section at the end of your article. And make sure you reference the work correctly—authors' names, date of publication, journal or text in which it was published, and the title of the article quoted. If two authors said much the same thing (Bloggs, 1992; Nurks, 1997) then separate the two using a semi-colon. Make sure both references are included in your reference list.

In the reference list, all references will appear in alphabetical order. If there are two or three references by the same author, they should be listed in date order.

The second system of referencing is the Vancouver system. This is used mainly for medical texts, and can drive a writer mad, and an editor insane. Why? Because they are listed by means of a number (actually, a superscript) in the order in which they are used in the text. So if you need to add another reference in paragraph one, then you will need to change all the references throughout the article, and the reference list too. In the Vancouver system, the titles of journals and books are treated differently from those in the Harvard system.

REFERENCE LIST

Lists of references form an important part of your article. They give credit to the authors whose work you have used, they enable a reader to check the source of your data, and they enable an interested reader to become even more acquainted with your subject.

Let's look at the Harvard system first. As mentioned, these are listed in alphabetical order. If it's a journal article, you will need the author and his or her initials, the date of publication (year), the title of the article, the title of the journal, its volume and issue number. And get the titles right. If it's a book, then you will need to quote the author, or authors, the date (year) of publication, the title of the book and its publisher and place of publication.

If quoting a journal article, this goes in apostrophes (' ... ') and the name of the journal goes in italics. If you are quoting a book, then the title of the book goes in italics.

Here are some examples.

Bloggs, G.J. 1998, 'Writing essays—the definitive student guide', *University Results* 67; 3-5.

Bloggs, G.J. 1998(b), 'Why students fail', *Curriculum* 33; 6-9.

Nurks, A.S. 1991, *Every Student's Guide to Essay Writing*, Doghouse Publishing, New York.

Now for the Vancouver system. As mentioned earlier, this is used mainly for medical referencing.

The order of things is different, and nothing at all goes in italics. And for these, each one will be numbered according to the order it was cited in your article. Nothing could be simpler, could it?

1. Bloggs GJ, Nurks BL, Writing medical essays—the definitive student guide to passing medical school, University Results 1998:56(6); 44-46.

2. Bloggs GJ, Why students fail, Medical Curriculum 1998:21(1);3-7.

3. Nurks AS, Every Student's Guide to Essay Writing, New York, Doghouse Publishing 1991.

There are numerous documents, such as those published on the Internet, that are not listed in this short list. Consult a style manual, such as the Scientific Style and Format, The CBE Manual for Authors, Editors and Publishers, latest edition. This excellent reference is published by Cambridge University Press.

Chapter 7 Tables, Graphs and Illustrations

Tables can present a lot of data in a simplified and compact way that, if prepared well, will make the reader's job of understanding your article easier. Tables should be submitted with the article but on separate sheets—one table to a page. They must be numbered in the order in which reference to them appears in the text.

Don't go to extremes in the use of tables. They should contain only data that is essential to the article, no more and no less. Too many tables in a journal article make for clumsy layout, and confuse the reader.

Each column of each table should be labelled with a brief description saying exactly what is in that column.

When referring to tables in the body text, include a reference to the effect TABLE 1 HERE. In the typography stage, the production staff will place the table as close to that point as possible, usually after its first mention in the text.

GRAPHS

Graphs can replace a lot of words. They show trends, they show performance, and much, much more. At least, a well prepared graph does. Graphs should never be cluttered so the data and trends are lost. They should never be so crowded that nothing can be interpreted from them. With bar charts, for example, it is possible to shade or colour bars so the interpretation is clear at a glance. And graphs must be easily interpreted with little concentration. Legends must be clear. Axis labels should be stated. The range of each axis should extend only slightly beyond the highest and lowest data. If the highest age group represented in a human population study is 58 years, there is little point in taking the Y axis to 200. If the last year of the study included data for 1998, don't take the X axis to the year 2050.

Be careful not to give false impressions through the use of graphs, either deliberately or inadvertently. If the range of variation of data is small, don't create a Y axis scale that ranges from 0.3 to 0.4. Only a small variation will show as a cataclysmic event. Make such a scale on the Y axis from 0 to 1.

When including graphs with your article, always include the data used to compile the graphs. Many publishers have a standard format for all graphs. Yours might not comply with that format, and it might be necessary for the production team to compile a new graph—they can do this only with your data.

PHOTOGRAPHS AND ILLUSTRATIONS

Photographs can be useful. Those depicting some technique, some non-human subject, are fine. But don't go down to a group of people and take a photograph just to illustrate your article. Such a photograph might not mean much on its own. Secondly, people don't always like being photographed. Thirdly, there can be all sorts of legal problems. Recently, a book was published about the immunization of children, so the authors went to a pre-school and took a photograph of some small children playing in the school grounds. Unfortunately, one of the children in the photograph (which was used on the front cover) was the centre of a custody dispute, and the father did not know the whereabouts of his child—until the book was published! So be warned. Be careful!

An illustration such as a diagram showing the main components of a plant or animal cell, or the layers of the atmosphere, can provide more information than can a lot of words. Use diagrams to aid clarity.

Chapter 8 The Last Bits

TITLE PAGE

There is a reason why the title page (the first page of your article) is discussed near the end of this book. It's one of the last pages you will compile.

The title page of the article you intend submitting to a journal should contain the title of the article, the author, or authors (in the correct agreed order—being first author, second author or so on is important to individuals as well as to their contribution). The names of the authors will be followed by their affiliations—the institution, such as research laboratory or university where they work. The address for the corresponding author should be clearly indicated (usually by saying 'corresponding author). This will be the person from whom readers will seek further information, or to whom the editor will address all questions. The title of the journal you are submitting the article to should be shown. There should also be the standard copyright sign, designated by the symbol © followed by the year of submission of the article.

ABBREVIATIONS

Each field of science has its list of commonly used abbreviations that are well known to readers in those fields. DNA, DDT and CSIRO are commonly used without the need to spell out in full what they stand for. Other abbreviations are well known—kph for speed, ml for volume, % as a portion, kg for weight. If you think your intended reader will not be familiar with an abbreviation, write it in full the first time, and type the abbreviation in brackets after the use of the word, so that the vehicles were timed at 100 kilometres per hour (kph) along the freeway.

Most abbreviations do not use periods (full stops) if the last letter of the abbreviation would normally form the last letter of the word, as in Dr for doctor, but Prof. would have a period.

But better still, don't use abbreviations at all—write kilometres instead of kms, professor instead of prof.

Chapter 9 Editing

Here we will look at the art of editing. Learn it, and apply it!

Once you have written the draft of your article, there remains a most important task ahead of you—editing it to take out all grammatical and spelling errors, ensure that the styles of all authors are consistent, and to make sure that no errors of fact remain. It is often this part of the writing process that many consider the most difficult, yet it is perhaps just as important as writing the draft. The standard of editing can often determine an article's acceptance or rejection by a journal.

Never expect to produce a finished text in one sitting. Consider the first draft as merely the starting point of what will become a good article.

Without a draft, you have nothing to revise. Without revision, your draft remains shapeless and incoherent. Both are necessary for clear, effective writing.

Be ruthless in striking out what is not necessary. A large part of revision involves chipping away excess words. Look for clarity. Strengthen important points by expressing them in short sentences.

Eliminate unsupported generalisations. Revise awkward repetitions of the same word, for example, 'such'. Replace vague abstracts with precise words having richer meanings.

Be alert for errors in grammar and usage, and in spelling. Make sure your punctuation is adequate and conventional but no more frequent than clarity or emphasis requires.

Beware of mannerisms of style, such as beginning too many words with 'but' and 'and'. Avoid writing long, complicated sentences, of several different points joined together with any number of ands and buts.

In editing, you will get rid of what is redundant or irrelevant. The result should be a coherent whole. In all sections of your article, omit all trivial and tedious details that are not essential for completeness. Don't repeat yourself.

Editing

What follows is a long list of what you should eliminate from your article, and other points you should check.

Editing is your chance to:

- Ensure a logical progression of ideas, rearrange the sequence of topic headings, paragraphs, sentences and illustrations;

- Expand or reduce text to improve the quality of the article;

- Remove inappropriate words, vagueness, break up long sentences or paragraphs so the text is easier to follow;

- Check that the headings and sub-headings are still appropriate to what follows them;

- Check again the length of your article against the journal's word limit;

- Improve the quality of the article so it is of literary brilliance;

- If you make changes at the last minute to the text, ensure that appropriate changes are made to the abstract;

- Don't keep the article at its original length merely to make it appear a long composition. There is a saying that all writing improves in proportion to the amount of words left out. Once you have completed the first edit (which should be a heavy edit, eliminating all possible words you don't need), go through again and eliminate more that you don't need. Do this until it is impossible to delete even one more word. And then it should be all right;

- Make sure you keep sentences short—about fifteen words on average is an ideal number to aim for, but vary their length for interest. Don't join a number of sentences together with 'and';

- Use simple words rather than complex words—use words of two syllables rather than five syllables;

- Use familiar words rather than those for which readers will have to consult a dictionary;

- Avoid unnecessary words;

- Ask yourself if you have considered your readers' background;

- Remove vague, meaningless and wrong words—replace these with concise words and phrases;

- Jargon—replace this with simple, precise terms or clearly defined technical terms;

- Remove all redundancies;

- Remove tautologies, clichés;

- Convert the passive voice to the active voice where possible.

We'll look at some of these points in more detail.

WORDS THAT DON'T CONTRIBUTE

Remove all redundant words, such as really, actually, basically, physical location, do in fact, at this point in time. Make sure every sentence is worded correctly. Make sure every word is the right word. And don't make up words.

CLICHÉS

Clichés are dull and unoriginal. Examples of clichés include: at this point in time; on the back burner; level playing field; cool, calm and collected; history tells us; the bottom line. Don't use any of them.

COLLOQUIALISMS AND SLANG

A colloquialism is a word or expression appropriate in conversation. It is out of place in technical writing, for example we have a swell program; square (meaning old-fashioned); cool. Don't use colloquialisms such as the satellite was rocketed into orbit. Such expressions are fine when you are writing for people who talk like that, but they are not appropriate for scientific or technical writing. This comes back to what we mentioned earlier about considering your readers.

WORDS TO USE

A good writer does not use words beyond the capacity of the readers. If you must use an unusual word, define it where appropriate; for example, in writing for mountaineers, the word crampon is all right to use. In an article for a different reader, this word would possibly be meaningless and would need defining. It is unnecessary to define any word that is, or should be, within the general knowledge of the readership. Ask yourself if a definition is necessary.

Check words that sound alike but are quite different, such as waste and waist, right and rite, straight and strait. Also, metre and meter—an engineer checks his 10 meters, while a child crawls 10 metres.

SIMPLE WORDS

Always use simple words instead of long, complicated words;

proceeded—went;

demonstrated—showed;

narrated—told;

make a recommendation that—recommend;

effect a reduction—reduce;

create an improvement in—improve.

UNNECESSARY WORDS

These are words that fill no function, convey no meaning and contribute in no significant way to the reader's understanding of the article. In other words, don't say in ten words what can be said adequately in three. Many writers string together too many unnecessary words, such as I was going to go tomorrow; the current situation should serve to start many people thinking. Use the minimum number of words possible; for example, words can be deleted from these examples:

twenty metres [away] from the creek;

we [ourselves] ...;

our [own] ...;

such as ... [for example] ...;

look [out] for opportunities;

the [biological] scientist;

on the basis of (by);

as a result of the fact that (because);

the way in which (how);

people who enter university to study for their degree (undergraduates);

emerged victorious (won);

had an effect on (affected);

have to [have a knowledge of] (know);

tends to be (is);

finally, [the last point];

so far as is [presently] known

bisect [in half];

modern life [of today];

[vital] essentials;

[sufficiently] satisfied;

it is [clearly] evident the patient hanged himself, [thereby taking his own life].

The phrase 'or something' is overused in speech, but means nothing at all (it is probably a shortened version of 'something like what I have said', or 'something similar to this'). This should never be used in writing. Redundant words include outside of the city, front side, left side, right side, (and definitely not backside!), walking on foot.

HYPHENS

Don't hyphenate words in your article, even if the word is too long to fit on one line. The hyphen could be inadvertently retyped and appear in the wrong place in the printed text. Better still, try using shorter words. Hyphens are acceptable, however, where there is a clear confusion. Compare recover (get better) and re-cover (as in a chair), recreation (a holiday) and re-creation (to create a second time).

EMPHASIS

Use italics instead of underlining words such as plant names, but don't overuse emphasis, such as italics, underlining or bold. Such emphasis should be used sparingly.

CHECK THE SPELLING

Check all spelling, particularly words that end with suffixes such as able/ible, en/ern.

AMBIGUITIES

Always make sure that no confusion appears in your sentences, for example 'our study found that children often anger parents. They won't talk to them' (who won't talk to whom?)

OBVIOUS BY IMPLICATION

Don't spell out ideas that are clearly implied. For example, the bacterium [is a microorganism]; over a long period [of time].

PHRASES TO AVOID

Don't write 'between June and July ...' There is no time between these months.

There is no time 'between 1998 and 1999' either.

Reduced by twice that amount ...' isn't very clear, so don't say it like that.

Three metres square is not the same as three square metres.

If your data range from 300 to 600, write 300–600, not 3-600. The latter means from three units to six hundred units.

An area is 100 mm x 200 mm, not 100 x 200 mm.

Vagaries such as 'some', a 'certain amount' and 'a number' are so meaningless they are best left to conversational English.

'Soon ...' How soon? Time is relative, we are told.

Be careful when talking about reduced temperatures. Twenty degrees is not half of 40 degrees (we need to consider the absolute temperature for such a statement to have any meaning at all, then it becomes too cumbersome. Just say that the temperature fell from 40 degrees to 20 degrees. The statement 'the temperature was reduced by three times that amount ...' is so bad it does not require further comment.

If you want other examples of meaninglessness, just analyse the radio and television news for an evening. These, unfortunately, are examples of what frequently is written in other forms of communication, but hopefully not in scientific articles.

Now see what you can do with the article you have written. Rearrange your work after you have cut out all unnecessary words. Ask: can I amalgamate any paragraphs when I rearrange them? Can I then cut out more words?

Chapter 10 The Final Check

The final check is possibly one of the most important checks you can perform on your article before submitting it to a journal.

In the final check, it is essential to check your references. As you go through the text, check that each publication and author is included in your list of references. Be particular about the order of the names of the authors and the year of publication. For example, the text might refer to an article by Bloggs and Nerd 1997. In the references, it might refer to Nerds and Bloggs 1996. Are they really the same article, or different articles by the same authors?

In the final edit, check for correct abbreviations, units of measurement, and order of tables and figures. If you move a table or figure further up or down the article, make sure that you renumber it and subsequent illustrations or tables if necessary, to make sure one has not been placed ahead of another numbered illustration or table. All numbering should be sequential.

Check the captions of the table and figures. Make sure they refer to the correct data. A frequent mistake is a caption that reads, for example, 'Measles outbreaks, Adelaide, July 1996–June 1997'. The graph might show data from January 1996, or to December 1997.

Check several articles published in the journal you are submitting your article to. Check their use of capitalisation and italics.

AUTHOR'S CHECKLIST

An article can change from what the writer intended. This happens more so when several authors are involved, each with different ideas about how the article should progress. So before submitting your manuscript, go through the final checklist. Ask yourself some questions.

- Does the article still fall within the scope of the intended journal?

- Is the subject still important to justify publication in this journal? (If an article is many months in the writing stage, a lot can happen — for example, a journal can publish so many articles on a similar subject, the editor and readers are rather tired of the subject). Your article might even be out of date.

- Does the abstract summarise the article? Is it effective?

- Is the article technically sound (no glaring mistakes, nor mistakes of fact).

- Is the literature review extensive enough? Quoting only one reference is not enough for a well-researched article.

- Is each section — particularly the Introduction, Materials, Methods, Results, Discussion and Conclusion — complete and in a logical order?

- Is the Methods section detailed enough for colleagues to repeat the study?

- Are the results clear?

- Is the Discussion clear and well reasoned?

- Is the conclusion the logical one to draw, or have I missed something? Do the results really mean something I haven't mentioned?

- Have I written the article well, free from all ambiguities, poor sentence and paragraph construction, punctuation, spelling?

- Does the article conform to the journal's preferred style?

Here is a list of common errors that creep into technical writing:

- Exaggeration of facts;

- Misinterpretation of data, often arising from omission of facts;

- Errors in data or terms;

- Conclusions based on faulty or insufficient evidence;

- Unreliable statistical methods and manipulation of data;

- Failure to distinguish between fact and fiction;

- Inconsistencies and contradictions;

- Omission of important topics;

- Incorrect order of sections or paragraphs;

- Placement of material in an inappropriate section;

- Incomplete development of a topic;

- Weak beginnings to sections;

The Final Check

- Inclusion of irrelevant, trivial and tedious detail;

- Style that is hard to read;

- Inadequate emphasis on interpretation and conclusions;

- Author does not explain why he or she is writing the article;

- Not clear where the article is going, and why.

Chapter 11 You Are Almost There

SUBMITTING YOUR MANUSCRIPT

All journals require articles to be submitted in a uniform standard. Such requirements include:

- Double spacing of all text;
- Each section (Abstract, Introduction, Methods, Materials, Results, Discussion, Acknowledgements and References) should begin on a new page.

Ascertain if the journal wants a disk provided at the time of submission of the article. Because the article is likely to be substantially rewritten after it has been sent out for peer review, many editors prefer a disk version after all the changes have been made. Others, however, insist on the disk version at the time of the original submission (which can create problems if the wrong disk is used).

It is important to ensure there is only one version of your article on the disk as this eliminates the chance of the wrong version going to print. If you can, save the article in a format recommended by the journal, otherwise save the text as plain text or rich text format (RTF), both of which are easy to read. If you have tables and graphs, save these onto the same disk.

Always, without exception, submit a hard copy (printed version) of your article with the disk. It is surprising how many authors submit only the disk with no instructions. It is not always easy to open files in a system that is not compatible, and there is a likelihood that data can be transposed (happens often) or is deleted (frequently occurs). A hard copy will make it easy to check that the disk version is complete, and is readable. The editor will work from the hard copy, not from the disk version. A note written on the disk about the system or program used to save the file will be very useful (Fodder trees for livestock production – GJ Bloggs. Macintosh Word).

The order of the pages should be:

Title page

Abstract and key words

Introduction

Methods

Materials

Results

Discussion

Conclusion

Acknowledgements

References

Illustrations, tables, graphs—one to each page

Captions

Chapter 12 The Revision

The revision of your article is going to make just that last bit of difference.

You might be asked to revise your article. This is not a rejection. It means that you are almost there, and there is a good chance that your article will be published.

If you are asked to revise it, it could mean that the reviewers' comments were minor, but you will need to address issues the reviewers raised. You might be asked to explain some method you used.

Perhaps you will need more emphasis on some points. But at this stage, you will have definite areas to fix up, so they should be minor.

Do this job urgently and get your manuscript back to the editor within a couple of days.

Occasionally, it will not be possible to agree with a reviewer's comments or suggestions. If you cannot accept those comments, revise what you can, and explain to the editor why the other points have not been addressed. A compromise can be worked out in many cases.

THE ACCEPTANCE

Congratulations! You have had your manuscript accepted by the journal! Acceptance does not mean that it will be published without any changes. Editors have a right to change what they consider needs fixing—style, grammar, tone, and numerous other details. But if your article has been accepted, you will be notified, and advised of its publication date.

CHECKING THE PROOFS

Editors will return page proofs for checking. These are copies of the pages that will appear in the journal. This will be your last chance to make sure everything is as it should be. Don't go changing anything, unless something is very wrong. Just mark errors; the editor will decide whether or not anything warrants changing. The copy that you sent to the journal should have been your 'final'.

Learn a lesson from this for next time! It is tempting to make changes, but don't. Editors may see confusion whereas authors don't. A recent example I encountered was with the term flying fox. In Australia, this term refers to the fruit bat. Overseas readers could imagine a fox with wings. The editors were not happy with the term flying fox, while the authors were not happy with fruit bats—as they pointed out, the fruit bats were more closely related to humans and other primates than they were to the true bats. Just accept that on occasions, no matter how strong your argument, editors win!

When checking your final page proofs, pay particular attention to ensure the following are correct:

- Title;
- Spelling of authors' names;
- The name of your institution or affiliation, and those of your co-authors;
- All figures and tables are included;
- Figures and tables are numbered correctly (check that none has been transposed);
- Captions are correct;

- All references are correct.

There you have it—just about everything you will need to know to write for refereed scientific and technical journals.

Good luck!

Index

Index

Index

www.ingramcontent.com/pod-product-compliance
Lightning Source LLC
Chambersburg PA
CBHW060637210326
41520CB00010B/1642